Lost in Transmedia

Exploring Fandom, Transmedia, and the Birth of Electronic Anthropology

---

CD Damitio

Indignified Press
2026

ISBN: 978-1-939827-58-6

Published by Indignified Press

indignified.com/books

For the fans who figured it out first.

Contents

Foreword: The Island That Predicted the Future 1
Chapter One: The Van, the Show, and Me 3
Chapter Two: A Three-Hundred-Year Island Story 6
Chapter Three: Engineering Uncertainty 10
Chapter Four: Who Watches, Who Obsesses 14
Chapter Five: Field Notes from the Island 18
Chapter Six: The Spoiler Wars 22
Chapter Seven: The Voices 25
Chapter Eight: The Third World 32
Chapter Nine: What Lost Started 35
Bibliography and Further Reading 39

Foreword: The Island That Predicted the Future

In 2008, I was a twenty-nine-year-old anthropology student at the University of Hawaii writing a thesis about a television show. My committee thought it was a slightly undignified subject. My advisor, to his credit, did not. He saw what I was trying to argue: that the show called *Lost* was not just a television show, and that the community forming around it was not just a fan club. Something genuinely new was happening, and it was happening on the island where I lived, involving a cast and crew I kept running into at grocery stores and car washes and casting calls in elementary school cafeterias.

That thesis became this book. I have expanded the arguments, deepened the theoretical grounding, and added the perspective of nearly two decades of watching everything I observed in 2008 become standard practice across the entertainment industry. But the core observation remains exactly what it was when I handed in my thesis on November 26, 2008: *Lost* was not doing something unique to *Lost*. It was doing something that would eventually be done by everyone.

The show ended in 2010. The finale was divisive in the way that only something deeply loved can be divisive. Fans who had spent six years mapping every Easter egg, decoding every anagram, building elaborate theories about the Dharma Initiative and the smoke monster and the numbers 4, 8, 15, 16, 23, 42 — those fans felt, with some justification, that the ending had not honored what they had invested. Other fans thought it was beautiful. Both groups were right, and the argument between them is itself a chapter in the story of what *Lost* started.

What *Lost* started is this: the systematic blurring of the boundary between the production and consumption of entertainment. The idea that a story can exist simultaneously as a television series, a website, a real-world puzzle, a community practice, a set of objects you can physically hold, and an ongoing collaborative enterprise between the people who made it and the people who watch it. In 2004, this felt experimental. By 2026, it is the blueprint for every major franchise in the world.

Marvel's connected cinematic universe. HBO's game of whispers around its prestige dramas. The *Stranger Things* season-to-season ARG. The metatextual horror of *House of the Dragon* fan wikis growing faster than the show. Streaming platforms engineering completion anxiety, recommendation loops, and parasocial attachment using the same architecture of contingency that Damon Lindelof and Carlton Cuse deployed in 2004 with a polar bear on a tropical island.

I got some things wrong in 2008. I got more things right than I expected. What surprises me most, looking back, is not what has changed but what has not. Fans still gather around texts they love with the same hunger, the same need for community, the same drive to find meaning in the details. What has changed is the platform. The Tropical Smoothie Cafe on Oahu where forty fans gathered to watch a premiere together — that ritual is largely gone, replaced by millions of people watching alone, connected only through social media feeds that were themselves engineered by the same architects of contingency I was studying.

The island predicted the future. I was just taking notes.

CD Damitio

2026

Chapter One: The Van, the Show, and Me

The first time I heard about *Lost* was on a Craigslist help-wanted posting.

It was 2004. I was twenty-five years old, living in my van in the parking lot of a KMart in Honolulu, Hawaii. This was not the romantic van-life of Instagram reels that would not exist for another decade. It was a practical decision: I had nearly finished my first novel and wanted to complete it without worrying about rent. The van served as shelter, office, and address. For mail I used a post office box in Kailua on the windward side of Oahu, which was also where the production office for the show in question was located, and which was, inconveniently, on the opposite side of the island from where I parked to sleep.

The listing said that a new show being produced on Oahu was seeking an office production assistant. I had been picking up work as a production assistant on visiting shows — *America's Next Top Model, The Apprentice* — so this seemed like the right kind of job. I put on clothes that did not smell like van and drove to the interview.

The girl at the front desk told me the show was like a cross between *Survivor* and *The X-Files*. That sounded like just the kind of show I would have loved to work on. The interview went well. My resume was solid, my enthusiasm was genuine, and I felt a good rapport with the producer, a man named Steve. He shook my hand and told me it was a decision between me and a woman who had interviewed earlier. I walked back to where my van was parked three blocks away and waited for a call that did not come that day.

Steve called the next day with the nicest rejection I have ever received. They had gone with the other candidate because she lived closer to the production office. It was completely reasonable. It was probably true. But I could not help wondering whether someone had watched me walk to the KMart parking lot and climb into what was obviously, to anyone paying attention, the vehicle in which I was also living.

I took a job driving limousines, found an apartment, kept working on the novel. And I watched *Lost* become an institution.

It is difficult, from 2026, to convey what it meant to live on Oahu during the years when *Lost* was being made there. The show was shot entirely on the island — its jungles, its beaches, its warehouses and neighborhood streets repurposed as Sydney and London and the Iraqi desert and a mysterious island somewhere in the Pacific. You could not go to the grocery store without running into someone who had worked on it. You could not drive certain roads without passing a filming location you had seen on screen. The show employed hundreds of local crew members, cast local actors as recurring extras, and used the Hawaii Film Studios at Diamond Head as a base of operations. *Lost* was not a show that happened to be filmed in Hawaii. For four years, *Lost* was Hawaii, or at least a substantial part of the conversation the island had with itself.

On August 28, 2004 — nearly a month before the broadcast premiere — the semi-weekly Sunset on the Beach event in Waikiki hosted the world premiere of the *Lost* pilot. More than ten thousand people came. Fourteen cast members were present. I was in the crowd, still slightly bitter about the job I had not gotten, watching a show I had already decided I would probably like. Thirty minutes in, the woman standing next to me turned with a look of complete shock and said, "Was that a polar bear?"

It was. We were all blown away. We knew we were seeing something different.

Four years later, completing a degree in Anthropology as an Honors Student at the University of Hawaii at Manoa, I was required to write a thesis. The question of what to write it about was, for me, not a question. For four years I had been watching something happen on the island where I lived — a new form of entertainment assembling itself from old parts, a new kind of fan community growing up around it, a new relationship forming between the people who made stories and the people who consumed them. I had access that no researcher in another city could have. I had been to the premiere. I had applied for a job on the show. I drove its cast members in limousines. I stood in line for three hours at an open casting call alongside fifteen hundred other people who all had the same improbable hope.

This is the record of what I found. It is also, inevitably, the record of a specific moment: 2008, the fourth season of *Lost*, the year before the final two seasons would play out and the island would offer its answer to every question it had spent six years asking. But ethnography is always a record of a specific moment. The data does not expire; it transforms into history. What I observed in 2008 about the relationship between a text, its creators, and its audience turned out to be the shape of things to come for the entire entertainment industry. That is worth recording.

One more thing about the van. I finished the novel. Nobody published it. Years later I published it myself, which is a different chapter entirely. But I want to be clear that living in a van to write a novel while also working as a production assistant was not, at the time, unusual on Oahu. The island attracted people who were doing things in unconventional ways. *Lost* itself was being made in an unconventional way, in an unconventional place, by people who were making it up as they went and, paradoxically, insisting that they knew exactly where it was going. I fit right in.

Chapter Two: A Three-Hundred-Year Island Story

Before *Lost* became a television show, it was a library.

The writers — led by executive producers and showrunners Damon Lindelof and Carlton Cuse — were avid readers whose tastes overlapped in specific and traceable ways. "Reading those books as kids had a real indelible impression on us as storytellers," Lindelof told me in October 2008, talking about the shared literary vocabulary he and Cuse brought to the writers' room. When one of them referenced a scene from *The Stand* or a moment from *The Magician's Nephew*, the other knew exactly what they meant. This shared reading history was not incidental to *Lost*. It was architectural.

Tracing the genealogy of *Lost* means ranging across three centuries of storytelling, from the first novel in the English language to the most current trends in internet communities. Understanding where the show came from is essential to understanding what it did and why it worked. The show did not invent any of its constituent elements. What it did was combine them, at the right technological moment, into something that had not previously existed.

The story begins in 1719 with Daniel Defoe's *Robinson Crusoe*, considered by many scholars to be the first novel written in the English language. The parallels to *Lost* are structural and acknowledged. Castaways on a mysterious island. Encounters with "others" — indigenous or mysterious presences. Moral, economic, and religious decisions forced by circumstance. The island as a crucible for character. Lindelof and Cuse have stated directly that the island survival tradition, beginning with Crusoe, was the foundation on which they built.

Jules Verne's *The Mysterious Island*, published in 1874, gets closer. Verne's castaways arrive from the sky (in a balloon). They experience strange phenomena. There is a dog, a submarine, a mysterious beacon from another castaway, supplies that appear from an unknown source. The creators of *Lost* have acknowledged this as a direct starting point. The mysterious island with a hidden purpose, populated by forces not immediately understood — this is Verne's structure mapped directly onto Abrams and Lindelof's pilot.

Lewis Carroll's work contributes the epistemological framework. Many *Lost* episode titles reference *Alice's Adventures in Wonderland* (1865). A prominent location on the show is called "The Looking Glass." Like Alice, the castaways have gone down a rabbit hole into a world that does not obey the rules of the world they left. Like Carroll's readers, the audience of *Lost* is maintained in a state of productive uncertainty. The rabbit hole is not just a metaphor; it is the literal structure of the show's narrative logic.

Moving forward in time, *Lost* draws from the serial fiction of early radio and film. The serial format — recurring characters, cliff-hanger endings, episode recaps, story arcs that unfold over weeks and months — was the dominant form of popular entertainment in the first half of the twentieth century. Radio serials like *The Shadow* and *Flash Gordon* trained an audience to return week after week, to carry the narrative in their heads between installments, to speculate about what would happen next.

The sociologist Joan Reitz defines a serial as "any medium issued under the same title in a succession of discrete parts, usually numbered or dated and appearing at regular or irregular intervals with no predetermined conclusion." The question of whether *Lost* had a predetermined conclusion was one of the most contested issues among its fans for all six seasons of its run. The producers always insisted it did. Many fans, and some close observers, were not so sure.

The soap opera tradition added another layer. Anthropologist Michael James Intintoli, studying the soap opera *Guiding Light*, described a narrative mechanism that sounds exactly like a description of *Lost*: "The unpredictability of an expected resolution is an extremely important factor in the creation and uses of suspense and anticipation." Soap operas taught viewers to identify with characters over long time periods, to invest in relationships and outcomes, to accept death and reversal as narrative tools rather than conclusions. *Lost* deployed every one of these mechanisms, then added others.

Dallas (1978) contributed the season-ending cliffhanger as a cultural institution. The "Who shot J.R.?" mystery of 1980 was, in retrospect, a prototype for the kind of inter-season speculation that *Lost* would industrialize. Between seasons of *Dallas,* the question of the shooter became a genuine cultural event. Between seasons of *Lost,* that question would multiply into hundreds of unresolved threads, each generating its own community of speculators.

The immediate predecessors to *Lost* in the lineage of cult serialized drama are easy to identify: *Star Trek: The Next Generation*, David Lynch's *Twin Peaks*, Joss Whedon's *Buffy the Vampire Slayer*, and J.J. Abrams' own previous show, *Alias.* Each of these built what media scholar Henry Jenkins calls a "discriminating and productive audience" — viewers who did not merely watch but analyzed, theorized, created, and organized.

Twin Peaks is the most direct ancestor. Lynch's show pioneered the serial mystery with no reliable resolution, the deliberate withholding of narrative satisfaction as a technique rather than a failure, and the cultivation of an audience willing to do interpretive work between episodes. *Twin Peaks* also, notably, was marketed in what we might now call a proto-transmedia fashion: the tie-in books, the character diaries, the "Laura Palmer's Secret Diary" published as a real object in the world. Before the internet existed to support it, Lynch was already gesturing toward the ecosystem approach to storytelling.

Buffy the Vampire Slayer added the fan community dimension. Whedon's shows built fanbases that created extensive secondary literature — fan fiction, analysis, argument — and in doing so extended the life of the text far beyond its broadcast episodes. Many of the writers and directors who would work on *Lost* came from this tradition. They knew what an engaged, creative fan community looked like. They had been part of one.

What *Lost* did with this inheritance was systematic rather than accidental. Lindelof and Cuse were aware of their predecessors. They were also aware that the internet had changed the conditions of reception in ways that made the cult television model of the 1990s obsolete. The internet made the audience's interpretive activity visible to the producers in real time. It made fan communities global rather than local. It made the transmission of information — including spoilers, speculation, and the producers' own communications — instantaneous. These changed conditions required a new model. *Lost* was that model.

When Lindelof and Cuse described their literary influences in the interview I conducted with them, what struck me was not the specific titles — King, Verne, Lewis, Rowling — but the function those titles served. "Reading those books as kids had a real indelible impression on us as storytellers," Lindelof said. The shared library created what he called "a shorthand for the story." When one could reference *The Stand* or *The Magician's Nephew* in the writers' room and have the reference immediately understood, it enabled a speed and density of communication that shaped what the show became.

This is worth naming: *Lost* was built on intertextual intelligence. The Easter eggs scattered through the show — the books on characters' shelves, the names of philosophers given to characters, the literary titles of episodes — were not decorations. They were structural elements, chosen from a shared library that both the writers and the most dedicated fans inhabited. The show rewarded viewers who had read the same books. It created a community of shared literary knowledge as one of the currencies of fandom.

Three hundred years of island stories, serialized narrative, and cult television, combined at the precise moment when the internet could host the global fan community they required. That is what *Lost* inherited. What it did with the inheritance is the subject of the next chapter.

Chapter Three: Engineering Uncertainty

Damon Lindelof and Carlton Cuse did not stumble into transmedia storytelling. They engineered it.

"When we first started doing the show we were immediately aware that because of the nature in which we were going to reveal our mysteries," Lindelof told me, "the audience was looking very closely at the show for all of these little hints or Easter eggs that we started throwing in." The key word is *aware*. The architectural decision to build contingency into every layer of the text was made before the first episode aired. The question was not whether the fans would analyze the show in detail. The question was how to reward that analysis in ways that deepened engagement without satisfying it.

This distinction — rewarding without satisfying — is the central engineering principle of the *Lost* ecosystem. A show that satisfies its audience's questions teaches its audience to stop asking questions. A show that rewards the act of questioning, while perpetually deferring the final answer, teaches its audience to ask more. *Lost* was designed for perpetual deferral. Every answer opened new questions. Every revelation created new uncertainty. The machine was built to run forever, and then, crucially, it was given an ending.

Lindelof and Cuse identified, both in their production decisions and in my interview with them, four distinct levels of engagement that the show needed to serve simultaneously.

The first level is the *casual viewer*. These are people who watch the show regularly but may miss episodes. The show needs to make narrative sense at this level, with accessible recaps and self-contained episode-level stories. If *Lost* only functioned for people who had seen every episode in order, it could not grow its audience.

The second level is the *dedicated viewer*. These fans miss nothing. They know the continuity, track the character arcs, and derive their primary satisfaction from the ongoing serialized narrative. The show's weekly cliffhangers and season-ending revelations are designed specifically for them.

The third level is the *group viewer*. These are fans motivated enough by the show's mysteries and events to discuss, speculate, and rewatch episodes with friends, family, and colleagues. Group viewers spread the show virally through personal recommendation. They are the engine of word-of-mouth. When my survey respondents told me that within days of each new episode 78 percent of them discussed it face to face with another person, and 97 percent said they speculated about the show's mysteries in those discussions — that is group viewer behavior at scale.

The fourth level — the one that required the most deliberate engineering — is the *transmedia fan*. This is the person willing to record the show and examine interesting scenes frame by frame. Who analyzes the names of characters and in-show businesses for anagrams (fans quickly found that "Mittelos," a bioscience company mentioned in season three, unscrambled to "Lost Time"). Who participates in alternate reality games, visits fake websites, and is willing to chase a story that extends into the physical world — across billboards, in bookstores, on phone calls.

The transmedia fan requires more than the past, present, and future presented in the course of the television narrative. They must be treated to a meta-story that exists independently of the TV show itself. Designing a single text that simultaneously satisfies all four of these engagement levels, without alienating any of them, is an extraordinary engineering challenge. *Lost* solved it more successfully than anything that had come before.

The term "Easter egg" — a hidden element within a text that rewards attentive viewers with additional meaning — existed before *Lost*. DVD extras, video game secrets, and film in-jokes were all forms of Easter eggs. What *Lost* did was systematize the Easter egg as a primary structural element of a broadcast television show.

The Dharma Initiative logo printed on a shark for less than one second of screen time. Sub-audible whispers embedded in the soundtrack. Reversed speech hidden in the audio. Split-second images. Historical significance in the names of characters (John Locke, Jean-Baptiste Emmanuel Zoe, Desmond David Hume). Books on characters' shelves that, read carefully, illuminate narrative intentions. Fake companies with websites in the real world. Billboards for a fictional airline. A published novel written by a fictional character from the show, sold in real bookstores.

Associate Producer Noreen O'Toole told me that the producers were "continually amazed at the speed with which the fans unearth even the most obscure Easter eggs." This amazement was also design. The Easter eggs were made discoverable because the point was not hiding them — the point was the discovery. Media scholar Henry Jenkins articulated, in his 1992 work *Textual Poachers*, the concept of cultural knowledge as "social currency" within fan communities. On *Lost*, the currency was specifically the knowledge of Easter eggs: finding them, interpreting them, sharing the interpretation, debating its correctness. The show manufactured currency for a community it was simultaneously building.

The ARG — alternate reality game — is a form of interactive narrative in which players receive clues through multiple real-world channels (websites, email, phone calls, physical objects) and must collaborate to advance the story. Before *Lost*, ARGs were associated primarily with marketing campaigns for films (the campaign for Steven Spielberg's *A.I.* in 2001 is frequently cited as an early example). *Lost* was the first broadcast television series to use ARGs as a primary story extension tool.

"The Lost Experience" ran between seasons two and three, revealing backstory for characters that would not appear in the broadcast episodes. The New York Times described it as "a multimedia treasure hunt that makes use of email messages, phone calls, commercials, billboards, and fake websites that are made to seem real." This was followed by "Find 815" between seasons three and four, and "The DHARMA Initiative Recruiting Project" between seasons four and five — the latter still in development at the time of my research.

"Find 815" included real Oceanic Air billboards appearing in New York City, directing fans to a fake airline website that led into the ARG. The producers released at least one novel — *Bad Twin*, a detective story written by "Gary Troup," a fictional passenger who died when flight 815 crashed — published by Hyperion Books with an account of how the fictional author had disappeared on a flight from Sydney to Los Angeles. Troup's manuscript was found, read, and burned by characters in the second season. The book appeared in real bookstores months later.

What the ARGs created was a space where the distinction between the fictional and the real broke down by design. This was not a glitch in the transmission. It was the point. Lindelof described it to me: "We wanted to write the show so that a mainstream audience could appreciate it, but if you were a rabid fan there would be more that you could find — and the best way to do that, because it is only forty-two minutes per episode, was to sort of go outside the show itself and start coming up with alternate ways, like on the internet, or as you call it transmedial storytelling."

The concept I returned to throughout my research — and that I believe best describes what *Lost* built — is what I called at the time the "architecture of contingency." The show was engineered to make its audience uncertain about the status of what they were experiencing. Is this clue real or decorative? Is this website official or fan-made? Is this spoiler genuine or a foiler planted by the producers? Is the island real, metaphorical, or something in between?

This uncertainty was productive. It kept fans engaged between episodes, between seasons, and even between the broadcast of the finale and the decade of discussion that followed. But it was also, in retrospect, a double-edged engineering decision. The architecture of contingency produced a community of extraordinary analytical creativity. It also produced, over six seasons, a set of expectations about resolution that no ending could satisfy. The fans who had mapped every Easter egg, traced every anagram, built the most sophisticated theories — those fans had the most to lose from an ending that chose emotional resolution over narrative completeness.

That tension between the producers' intention ("Carlton and I are just trying to tell the character stories") and the fans' expectation ("every detail must mean something") is not a failure of the show. It is the structural consequence of building the architecture of contingency so successfully. When you teach an audience to find meaning in everything, you cannot then tell them that some things had no meaning. The island created the expectations the island could not fulfill. That, too, was part of the inheritance.

Chapter Four: Who Watches, Who Obsesses

In April 2008 I sent a survey to several of the most popular *Lost*-focused message board communities. Thirty-two people responded. Twenty-six were female and six were male. All were over eighteen. The median age was 36.5. All but four respondents were American citizens; the others were from the Philippines, the United Kingdom, Germany, and Canada.

These numbers require careful interpretation. They may tell us that American women over thirty are the largest fans of *Lost*. Or they may tell us, equally plausibly, that American women over thirty are the most likely demographic to complete online surveys sent to fan message boards. I noted this uncertainty in my thesis and I note it again here: the demographic data is suggestive, not definitive. Larger-scale studies conducted during the same period found similar female skew — Jonathan Gray and Jason Mittell's 2006 spoiler study found 60 percent female respondents — but the precise shape of the *Lost* fandom was never fully mapped.

What the data reliably shows is the behavior. And the behavior is striking.

Ninety-one percent of my respondents said they never missed a broadcast episode. Sixty-two percent watched episodes online. Sixty-three percent watched on DVD. Seventy-eight percent watched episodes more than once. That same seventy-eight percent said the reason they rewatched was specifically to look for hidden visual or auditory clues — Easter eggs — they might have missed on first viewing. Nearly all of them frequented websites associated with the show. Sixty-six percent regularly listened to podcasts about it. Forty-seven percent played the ARGs.

These are not the viewing habits of a television audience. These are the research habits of a community engaged in ongoing collaborative interpretation. The text of *Lost* was not, for these fans, something they consumed once and moved on from. It was an object of sustained study, the way a religious text or a canonical work of literature is studied — returned to repeatedly, cross-referenced, annotated, argued over. The show had become, for its most engaged viewers, a practice.

One specific community I embedded myself in during my fieldwork was the Hawaii Lost Fan Club, an online group that had grown out of a message board at Hawaiilost.com in 2004 and migrated to Yahoo Groups in 2006. At the time of my research the group had 125 members, primarily but not exclusively based on Oahu.

The Hawaii Lost group illustrated one of the most interesting social dynamics of online fan communities: the translation of digital familiarity into physical warmth. When I met group members in person for the first time, they greeted me as someone already known. The months of online discussion — commenting on episodes, speculating about mysteries, sharing news of filming locations around the island — had created a form of relationship that transferred directly into real-world interaction.

This dynamic extended beyond the local membership. In February 2008, the public boards of the Hawaii Lost fan club showed the introduction of a forty-seven-year-old German tourist who went by "Netti." She had traveled from Germany specifically to visit filming locations on Oahu and meet up with group members she knew only from the internet. When I met her at the Tropical Smoothie Cafe, she was firmly integrated into a group of local women who had been escorting her around the island — showing her locations, taking her to dinner, offering her a place to stay. Someone who had never been to Hawaii, who spoke English as a second language, who had no prior personal connection to anyone on the island, had arrived to a pre-existing community that adopted her immediately because of a shared investment in a television show.

This is not a trivial observation. Fan communities had existed before the internet — the Trekkies, the *Buffy* forums, the soap opera message boards. But the internet had transformed their reach, permanence, and capacity for cross-cultural connection in ways that were, in 2008, still being understood. The Hawaii Lost fan club was a demonstration that the community created by transmedia engagement could function as genuine social infrastructure — providing hospitality, friendship, and belonging to strangers across national and cultural lines.

My most revealing piece of fieldwork came from comparing two venues where fans gathered to watch the fourth season premiere: Tropical Smoothie Cafe in Honolulu, and Pearl, a high-end ultra-lounge in the Ala Moana Shopping Center.

At Tropical Smoothie, nearly forty fans gathered. The event was emceed by a local blogger, hosted by Teddy Wells — who owned the cafe and had played a character in season three — and ran through the commercial breaks with ticket giveaways for *Lost* merchandise. The crowd was diverse: fans from Croatia, Norway, Korea, Hawaii, and the mainland United States. Ages ranged widely. People were casually dressed, a mix of couples who had left their kids at home for a rare evening out and younger fans who had organized through the online group. The atmosphere was communal, engaged, and slightly chaotic in the way of people who genuinely cared what happened next.

Pearl was different. It was upscale, hip, filled with singles and well-dressed couples in their twenties and thirties. I visited three times. On my first visit, no one except me was paying attention to the show. On my second visit, the volume on the televisions was lower than the ambient noise of the bar. On my third visit, I had to ask the bartender to turn the television to *Lost*. He told me he had seen a few episodes but wasn't impressed, and that I was the first person to ask him about it.

The contrast is illuminating. The fans at Tropical Smoothie were — in the demographic language of television — less desirable than the patrons of Pearl. They were less affluent, less conspicuously young, less fashionable. They were also, observably, more socially connected to each other through the show. "The Lost fans were more socially awkward than the singles at Pearl," I wrote at the time, "and there was always a sense that they were watching the show together but separately." Fans who had met online but didn't recognize each other in person sat next to each other without speaking.

This social awkwardness — the online community that was more comfortable as an online community even when assembled in the same room — was not a failure of the fan community. It was a characteristic of it. The shared text was the social infrastructure; physical proximity was secondary. The show did not require co-presence to create community. The community existed in the discussion, the speculation, the accumulated shared knowledge. The room full of people who weren't talking to each other were nonetheless deeply connected through a practice that the people at Pearl, however socially fluent, had never joined.

Chapter Five: Field Notes from the Island

Anthropology has a tradition of the participant-observer: the researcher who does not merely study a community from the outside but enters it, takes part in its practices, and uses that participation as a source of data. In studying the ecosystem of *Lost* from Oahu, I had advantages no researcher in another location could claim. I was already embedded in the production island. I had attempted to join the production team. I watched the premiere on the beach in Waikiki with ten thousand other people. I lived where the show was made.

This proximity produced data that surveys and online ethnography could not.

I spoke with five crew members from various departments of the *Lost* production. All agreed to speak only on condition of complete anonymity, concerned that anything they said might affect their ability to work on the show or other productions in the future. I will honor that condition here as I did in my original research.

The best things about working on *Lost*, according to my informants, varied by position. Those in unpaid internship roles through the Academy for Creative Media at the University of Hawaii praised the professionalism of the production and the access to equipment they could not otherwise handle. Those in paid positions listed, simply and directly, the rate of pay. One told me that the best thing about working on *Lost* was that there was always food available. "There is food everywhere on that set, man! It's amazing that all of those people aren't fat."

All informants valued having *Lost* on their resumes. Two students felt that the contacts they made on set were the most valuable long-term asset. Interestingly, only one of my five informants had seen the show before getting the opportunity to work on it. One told me he had watched two seasons in a single weekend to prepare.

The less comfortable data: one informant described a pattern of racist jokes and comments from some crew members, including one instance in which he was the direct target. Another informant mentioned homophobic comments made in a climate of apparent easy familiarity that nonetheless made the informant uncomfortable. A third described drug use among crew members during production — marijuana during the long periods of waiting that are standard on any set. None of these accounts described the production as hostile. They described it as a large, multi-department operation with the range of workplace cultures you would expect from any such operation.

What struck me most in the crew interviews was the near-universal report that nobody on the local crew talked about the show. "Nobody seemed to care about the show at all among the local crew," one informant told me. "I would try to talk to the crew and they just didn't care. It was just a job to those guys." The people most intimate with the physical production of *Lost* — who built the sets, drove the equipment trucks, operated the lights — were largely indifferent to it as a cultural object. The show's meaning was created entirely downstream, by the audience.

On August 15, 2008, the Honolulu Advertiser reported that *Lost* was holding an open casting call at Ala Wai Elementary School. I decided to go.

At the very least, I would meet fans while standing in line. At best — the article listed the day rate for background actors at $759 — I would get hired as a regular on the show. I was not the only person to have had this thought. The producers later said over fifteen hundred people came out that day.

I stood in line in the Hawaiian sun for more than three hours. In that time I spoke with perhaps a dozen people. All of them were fans of the show.

Directly ahead of me were three military wives and their children, having driven in from Schofield Barracks on the North Shore. One of the women had gotten obsessed with the show and pulled her two friends into it; when they saw that it might be possible to become actors on it, they loaded the kids in a car and drove to Honolulu to wait in the heat. The woman who had introduced the others to the show told me it was her daughter's second birthday. The daughter, roughly two years old, was remarkably well-behaved as we inched forward in the sun.

Ahead of the military families was a woman with her two teenage daughters. The mother's answer to why they watched the show was immediate and complete: "Because it is the best show on television." The older daughter added, without a beat, "And because of Desmond." Her sister and the military wives quickly agreed. The actor Henry Ian Cusick, who played Desmond, was a consistent presence in my research — generating write-in votes on my survey, generating enthusiasm at Tropical Smoothie, generating a disproportionate number of fan discussion posts that I am comfortable describing as not primarily about the plot.

Behind me were three generations of women — a grandmother, a mother, and a daughter — who had seen the casting call on Craigslist and come in the hope of seeing some of the actors. No actors were present. They left after a couple of hours, before even filling out the paperwork. The casting process, when I reached the cafeteria at the front of the line, consisted of having my picture taken and being rushed out the back door to make room for more. Everyone was in the same line, with the same improbable hope, baking in the same Hawaiian sun.

Living on Oahu while *Lost* was being made there meant regular incidental encounters with cast members in the circumstances of ordinary life. I want to record several of these because they reveal something about the particular texture of celebrity as it existed within the ecosystem of the show's production island.

I met Dominic Monaghan — who played Charlie, the washed-up rock star with a heroin problem — at the gym. We were both working out. There was not much conversation. I met Jorge Garcia — the real person known to millions as Hurley, the large, good-natured cursed man — at a supermarket in Kailua. At first I could not place where I knew him from. After an awkward moment of mutual assessment I said, "Oh, you're on *Lost*." He smiled and asked if I liked the show. I said I did. We moved on through the aisles.

The most interesting cast encounter came on Christmas Eve of 2005. I was still driving for a limousine company. I was sent to pick up a passenger in Lanikai, an upscale community on Oahu's windward side, traveling under the alias "Annie Hall." When she came to the door, I thought she was my boss's wife. They looked alike. This mistake produced a conversation in which I acted as though I knew her and she tried to figure out where she knew me from, since I was clearly not treating her as a famous person. It was only when her boyfriend — Dominic Monaghan — came to the door that I understood my error. She was Evangeline Lilly, who played Kate. She was catching a flight to Vancouver to spend Christmas with her family while he stayed in Hawaii.

During the forty-five-minute drive from Lanikai to the Honolulu airport we talked. She spoke about what it felt like to become suddenly famous — the strangeness of it, the way it changed how strangers related to her. She described the speculation among the cast members about where the show was going, since the writers guarded the script closely and the actors often knew little more than the audience did. We talked about books we had read. She was a very normal, genuinely pleasant person who happened to be playing one of the most-discussed characters on the most-discussed show on television.

At the airport I met Monaghan briefly as he helped her with her bags. He was friendly. She thanked me and got out. I drove back across the island thinking that if I had gotten that job in 2004, I might have spent the last year and a half working with these people instead of driving them to airports.

I also met Yunjin Kim — who played Sun — at a car wash. We sat in the waiting area while our cars were dried. We were having a pleasant conversation, and as she was leaving I asked her name. She told me. I realized immediately who she was, felt profoundly foolish for not recognizing her, and did not ask for her number as I had been planning to before I found out she was a major television star. I have regretted this on at least two occasions since.

I include these encounters not as celebrity name-dropping but as data. Living inside the ecosystem of a major show's production created a specific kind of friction between the fictional and the real that was unique to Oahu and that illuminated, from below, the same blurring of fiction and reality that the producers were engineering from above. I knew these people as my neighbors before I knew them as characters. That order of knowing changes things.

Chapter Six: The Spoiler Wars

A spoiler is any piece of information that reveals something about a narrative earlier than the normal linear experience of that narrative would provide it. This definition, simple in structure, conceals a social and psychological complexity that the fan community of *Lost* argued about with the intensity usually reserved for politics and theology.

In my survey, 62 percent of respondents admitted to reading spoilers. Thirty-eight percent said they did not. Within both groups there were gradations, exceptions, and notable amounts of apparent guilt.

"I do it because I am weak," one respondent wrote, of reading episode previews and set photos. "Yes, I can't resist!" wrote another. "Yes, need my fix. It doesn't ruin it for me." One respondent offered what I came to think of as the definitive statement of the spoiler reader: "I read spoilers. Evidently, I can't wait to find out what is going to happen next but I enjoy being kept guessing."

The self-medication metaphors run through the spoiler responses with a consistency that cannot be accidental. Fans describe their spoiler reading in the language of addiction: the fix, the weakness, the inability to resist, the knowledge that it is bad for them combined with the inability to stop. One respondent even wrote about spoilers "responsibly" — making a conscious effort to avoid "game-changing" spoilers while allowing themselves smaller ones, the way a person might have one drink but not five.

Jonathan Gray and Jason Mittell, studying *Lost* spoiler behavior in 2006, found that 69 percent of their respondents consumed spoilers to some degree. My 2008 study found 62 percent. The convergence is striking. Independent samples, two years apart, both find that roughly two thirds of the most engaged fans of the show were actively seeking information that they simultaneously described as diminishing their enjoyment of the show. This is not paradoxical once you understand what spoilers actually provide.

Spoilers are not primarily about knowing what happens. They are about managing anxiety.

The architecture of contingency that *Lost* built — the deliberate cultivation of uncertainty about characters, plot, and the nature of the island itself — created a sustained state of low-level suspense that persisted between episodes, between seasons, between the broadcast of each installment and the community discussion that followed it. For most fans, this suspense was pleasurable. It was the point. But for a substantial minority, the pleasure of suspense was indistinguishable from the discomfort of anxiety, and they managed that anxiety through information.

A spoiler that reveals a beloved character will survive the finale is not just information. It is reassurance. A spoiler about the next episode's structure allows the anxious viewer to prepare for what is coming rather than being blindsided. This is why fans who claimed not to want major spoilers would still read "minor" ones — episode titles, casting news, filming locations — while maintaining that they wanted to be surprised. The surprise they wanted was emotional impact. The information they sought was structural safety.

The producers of *Lost* understood this dynamic and responded to it with a counter-weapon: the foiler.

A foiler is an intentionally misleading piece of information presented as a spoiler. It has no basis in reality. Its purpose is to send the spoiler-reading community in the wrong direction, protecting genuine plot developments by flooding the information ecosystem with false ones. There was a general consensus among the fans I spoke with that the producers were active — through intermediaries, under pseudonyms, or through the strategic use of leaks — on the most popular fan message boards, and that not every apparent spoiler was what it claimed to be.

The foiler is a fascinating cultural artifact. It represents the producers actively colonizing the fan community's information infrastructure, using it as a tool for narrative control. The fans who sought spoilers were, in a meaningful sense, playing a game the producers had designed. Even the act of seeking information outside the text was, for the most sophisticated fans, understood as part of the text — a layer of the architecture of contingency that the producers had extended into the meta-narrative space of the fan community itself.

The event that changed the spoiler culture of *Lost* permanently was the leak, at the end of season three, of the complete synopsis of the season finale. This was what fans called a "game-changer" spoiler: not an episode title or a casting rumor, but the full arc of the most anticipated episode of the season. The leak changed how fans felt about spoilers, and it changed how the producers thought about protecting their story.

Several of my survey respondents cited this incident directly. "I used to read spoilers but after I found the full disclosure of the S03 finale it really showed me that the show is a lot more fun when I don't know what's happening ahead of time." Others noted that after the incident, they became more careful about distinguishing genuine spoilers from foilers, knowing that major leaks were both possible and rare.

Ryan Ozawa, the podcaster, was widely known among the fan community as a source of spoiler information — Lostpedia described him as "the spoiler king," though he disputed this characterization. When I asked him about the season three leak, he was characteristically elliptical: "Snake in the Mailbox?" he said. "Ah. Well, you have to go look for spoilers." When I pressed: "I didn't touch it. Most spoiler hounds I deal with didn't, either. There was a big spoiler ethics debate after that one, to be sure."

Spoiler ethics. The phrase might seem overblown. It is not. Within the fan community, questions about what information was acceptable to share, how to share it without spoiling people who did not want to be spoiled, how to distinguish genuine leaks from planted foilers, and whether consuming spoilers was a form of betrayal of the show's creative intent — these were genuine ethical debates, conducted with the seriousness appropriate to genuine ethical questions. The fans were not taking the show too seriously. They were taking their community's shared practice seriously. That is a meaningful distinction.

Gray and Mittell wrote, in their 2006 study, that the between-episode gaps for *Lost* fans "are not barren periods devoid of Lost, but serve as times to focus intently on the program — rewatching episodes, contributing to websites, writing fan fiction, and consuming paratexts." The spoiler debate was part of this ongoing engagement. The text of *Lost* never stopped. Only the broadcast schedule did.

Chapter Seven: The Voices

The most direct evidence I gathered in my research came not from surveys or observation but from four interviews conducted in the fall of 2008: with Ryan Ozawa, the producer of the most popular *Lost* podcast; with Melinda Hsu Taylor, a writer on the show; with executive producers Damon Lindelof and Carlton Cuse; and with associate producer Noreen O'Toole. These are primary sources with no equivalent elsewhere. I present them here substantially as they occurred, with light editing for clarity and to remove academic interview scaffolding that would interrupt the flow.

Interview conducted September 15, 2008, via Google Chat. Ryan Ozawa produces "The Transmission," one of the most widely listened-to podcasts about Lost, with approximately 14,000 listeners per episode.

CD: What got you involved in Lost in the first place?

RO: The buzz over the pilot. Ten million dollars for two hours? Had to see that. Also liked J.J.'s stuff, so I was curious.

CD: What about your podcast — what made you switch from your previous podcast to The Transmission?

RO: Didn't mean for the previous podcast to end, or at least go into hibernation, but it turned out that way.

CD: What's the most people you've had tune in to any given episode?

RO: Hard to say. Average first-week downloads are 14,000 or so. I think the biggest number I saw was 24,000, but that was over a longer period.

CD: Do you think Lost is unique in the world it has created — all of it, fans, mythology, show, online, podcasts, virals?

RO: It's up there. It's certainly Exhibit A in shows and brands that extended far beyond what shows on screen.

CD: Do you think it will reach the same sort of status as Star Trek?

RO: Hard to say. Times are different now, for one. I wonder how Star Trek would've fared in a wired world.

CD: Do you think people that watch Lost are different than people that don't?

RO: There's a science fiction aspect to Lost, and more so now, but it's not as fantastic as Trek in terms of scope and imagination. I don't know if we'll be as nostalgic about Lost as Trek because Lost is more conventional drama, just with a lot of mythology. And no — my mom watches Lost. She's not like me. She likes Monk and CSI Miami. Do I think the fans that follow Lost online, listen to podcasts, read blogs, join forums, do the ARG — are they different? Sure.

CD: How so?

RO: I'd say the vast, vast majority of Lost viewers don't dig any deeper than the cover stories in EW. But fandom online? That's another dimension and demographic. I don't think viewers of Lost, on balance, are different than other TV viewers — at least those with good taste.

CD: What is it about Lost that you love?

RO: I love the show. When the show challenged my love, that was when we stepped back. It's not unconditional, to be sure. In terms of reward, though, the fan interaction is 99 percent of it. What we saw at Comic-Con, the people we met, made all the late nights worth it.

Interview conducted October 16, 2008. Melinda Hsu Taylor joined the writing staff of Lost in July 2008.

CD: Has Lost changed you at all?

MHT: It has changed me professionally because it's a huge credit and it is a huge step up for me. It's great to be working on a show that I love and that I'm passionate about creatively, so it's positive forward motion. I've only been working there since July, so I think it's too early to say if it has fundamentally changed me as a person.

CD: Is there much interaction with the fans?

MHT: The other day I was walking through the building and noticed a care package sent to the show that was making a plea for the Jaters — that would be Jack and Kate as opposed to Skaters, which would be Sawyer and Kate. Included with it were photo-shopped pictures of the happy couple in different locations. The fan was just making the point that they should be together. But that's about it. I have friends that watch the show and we all talk about it in the writing room.

CD: Is there a specific demographic the show is trying to hit?

MHT: The demographic we do really well with is 18-49 both sexes, more educated than not. Those are the people that buy stuff, so it goes over well with the network if we do well with them. But I don't think we're actually shooting for a specific demographic. I would say that our fans are more educated than not, with a college degree of some sort.

CD: Do you or the other writers have anything to do with the transmedia storytelling?

MHT: Some of the writers participated in the Comic-Con video and the interactive games, but so far I haven't worked on any of the transmedia stuff yet. It's all collaborative though, so in a way everyone takes part.

CD: Is Lost trying to change the world?

MHT: Short answer, yes. Creating a show that has changed marketing, the online community, and inspired passionate followers. Carlton and Damon want people to consider certain questions — like in "Man of Science, Man of Faith," the conflict between Jack and Locke. The empirical versus faith. Locke is just going to push the button and he and Jack are going at it, and then finally Jack pushes the button. It's a question of where does faith end and begin. They want people to think about these things.

CD: Is it fun to write for Lost?

MHT: I think it's really fun. The desert island allows for a heightened kind of storytelling with basic frontier dynamics. It's about what you want and need in the moment and the trappings of real life. Everything is pretty well fleshed out in the writing room so that there isn't a whole lot of room to add things in. Sometimes though you can add in one of the numbers or something else that the fans are going to love.

Interview conducted October 20, 2008. Lindelof and Cuse co-created Lost with J.J. Abrams and have served as showrunners since the pilot. This interview was arranged through associate producer Noreen O'Toole.

CD: One of the things I love about Lost is its connection to literature. Was there a conscious decision to make the show appeal to readers?

DL: A lot of those influences are in the show. Carlton and I are both avid readers. I'd say that we both read our own things but there is clearly a place where our tastes overlap — so where Carlton is more of an Elmore

Leonard spy novel fan, I've read all the Harry Potter books as well as the Narnia books, and of course all the Stephen King books. We have this common language of literature that creates a shorthand for the story. Like, it's like this in *The Stand*. Or Carlton can say, remember the Magician's Nephew with those rings. So we kind of fall back on that stuff a lot, and obviously reading those books as kids had a real indelible impression on us as storytellers.

CD: The fans who responded to my survey are heavily women in their 30s, 40s, and 50s. Were you guys trying to make Lost a sort of soap opera?

CC: No, we don't think of things from a demographic standpoint. We try to make the show from an instinctual place. Our basic rule is that if both of us think it is cool, it goes in the show. We really rely on our guts. Who watches it is really out of our control. We just try to make the show that we ourselves would most like to watch, and that has guided us from the beginning.

CD: There's no doubt that Lost is transmedia. The world of the show crosses into all kinds of media. How did this happen?

DL: When we first started doing the show we were immediately aware that because of the nature in which we were going to reveal our mysteries, the audience was looking very closely at the show for all of these little hints or Easter eggs that we started throwing in. We wanted to write the show so that a mainstream audience could appreciate it, but if you were a rabid fan there would be more that you could find. And the best way to do that — because it is only forty-two minutes per episode — was to go outside the show itself and start coming up with alternate ways, like on the internet, or as you call it, transmedial storytelling. In terms of having seen it done before — like the message boards in the Whedonverse, which I was very active on — I think we had the idea that, not that we birthed the idea of ARG, certainly movies like A.I. were doing that long before us, but the idea of doing it for a TV show felt unique to us, and we had the idea, with a lot of support from the network, to put that idea out there.

CD: How much contact with the fans do you have?

CC: Honestly, we really have very little direct interaction with the fans. Other than fans submitting questions to us that we answer in our podcast, we don't troll the boards. In fact we kind of stay away from them

— because what happens is, one, if you read criticism, that criticism can really stick in your brain. A compliment rolls off your back but that piece of criticism starts gnawing at you. And two, the people on boards tend to be more mythologically oriented. I think the mythology is what everyone obsesses about and discusses, but for us it is really like the frosting on the cake. The cake is the character stories. We find it's important to write — I wouldn't say a soap opera, but — a real character-centric story. It goes back to the literary references. It's like *The Stand*, which is a thousand-plus pages unabridged. You've got this high-concept idea of a super flu that has wiped out most of the population, but what *The Stand* as a book is, it isn't the details of that super flu. It's about the characters who are survivors of it. We try to take the same approach with Lost.

CD: Anthropology has been defined as the recontextualization of meaning over time and space. Does that describe what Lost is doing?

CC: I don't think we intellectualize it like that. We just see ourselves as storytellers continuing a tradition of storytellers who have done the kinds of things we've done but in other mediums. The imaginations of Jules Verne or Stephen King or C.S. Lewis or even J.K. Rowling — we are just trying to take the experience we had as readers and do the same for the people who watch our show. I think it is easy to aggrandize what we do or over-intellectualize it. We just think more of it in the emotional terms of: will this be cool, will this be scary, will this make you laugh, will this make you cry. We just try to turn our ideas into emotional experiences.

DL: When you look across the cultural landscape of media — movies and TV shows and books that have some sort of cultural impact — we hope and fantasize that ten, fifteen, twenty years from now, Lost is still something that people talk about. I was reading an Entertainment Weekly article yesterday about *The Prisoner*, which just released its full sixteen-episode box set. It was only sixteen episodes done in the 1960s, but people are still talking about it. It was representative of a certain place and feeling — when feelings about secret agents and mistrust of government reached a peak, and that show was a kind of commentary about how Europe viewed the world. If that makes it an anthropological study, I'm not sure Patrick McGoohan had any idea, forty years later, we would still be talking about it in those terms. So you just never know. As Carlton said, we are just trying to do our best version of the show, and we will let everybody else figure out what they think it is. At the end of the day, the legacy of Lost is really determined by how well it ends — and we feel that our biggest contribution, as its stewards, was securing an ending instead of just going on and on until no one watched anymore.

CD: Are there specific messages or ideas you are trying to get fans to think about?

CC: We basically want people to think about some of the things that we think about as writers. The dramatic issues that interest us include the notion of faith versus empiricism. We live in a country which is so polarized along those two divides — a deeply faith-abiding country that is also a culture of science and empiricism — and as people we are looking for what is the balance and how do you find your place in the world between those two poles. This year we are exploring destiny versus free will — how much of our lives are affected by environment, how much by choice. These are the kinds of things that we talk about in the abstract and they percolate into the writing of the show. I would say there is an intentional ambiguousness in our storytelling which allows viewers to draw their own conclusions.

CD: Does Lost owe more to the Bible or to The Twilight Zone?

DL: That is an impossible question to answer because I'm sure Rod Serling was a big fan of the Bible — most of those stories are almost as weird or supernaturally driven as an episode of *The Twilight Zone*, just one is accepted as science fiction and one is accepted as the definitive religious writing of many cultures. To say that one has more sway than the other as an influence on the show would be a Rorschach test for whoever is watching it. So if you are a biblical scholar, you can watch the show and take away an extreme sense of all these things Carlton was discussing: destiny and purpose, whether Aaron should be baptized, the struggle of Mr. Eko with being a priest, Locke and Jack arguing whether they should push the button. They're not arguing biblically, but the argument basically boils down to whether or not Abraham should sacrifice Isaac. But if you are an atheist who loves science fiction and doesn't like any of that stuff, you can watch the show on that axis — and you will probably gravitate toward characters like Sawyer, who doesn't care about free will or faith or religion. He just cares about making out with Kate and stealing people's stuff. We look at it through the spectrum of the characters. If all of our characters were interested in the same thing, it wouldn't really be worth having fourteen of them.

Interview conducted November 4, 2008. O'Toole served as assistant to Damon Lindelof and as associate producer on Lost. The interview was abbreviated due to a family bereavement in Lindelof's family and concurrent production demands.

CD: Do you see Lost as different from other shows?

NO: Absolutely, on every level. Obviously there are certain shows it can be compared to — with *Twin Peaks* there is that idea of the original cult show, there is a certain formula it follows — but it's so outside the box. It blends in mythology, drama, and all the other factors. I've never seen anything else like this on television and I don't think we ever will.

CD: Does fan participation make a difference in the course of the show?

NO: It depends on if you are talking about the blogging or the marketing campaign. In some ways, absolutely the fan participation makes a difference. We love the fans. We've seen different versions of them. We've seen them at Comic-Con. We released the Dharma Initiative commercial around the last season finale and just to see the participation level of the fans — from the websites all the way to Comic-Con and how they followed each and every step of the way — it's truly amazing. It makes a huge difference to us that the fans are willing to participate in this way.

Chapter Eight: The Third World

After several months of fieldwork, I sat down to write the conclusion of my thesis with the working premise that I had disproven my initial hypothesis.

I had set out expecting to find something genuinely new. I found, instead, that *Lost* was descended from three centuries of island stories, serialized narrative, and cult television; that its fans behaved much as fans of other shows behaved; that its producers thought of themselves as storytellers in a long tradition rather than inventors of a new form. In its constituent elements, nothing about *Lost* was unique.

And yet something new had appeared. I had to account for it.

What I proposed calling the "Massive Online Televised Interactive Multiverse" — MOTIM — is the term I used to describe the specific combination of conditions that *Lost* instantiated in 2004 and that I observed in full operation in 2008. A MOTIM is not a show, not a fan community, not an ARG, not a wiki, not a podcast network. It is the specific configuration created when all of these exist simultaneously, in relation to each other, generating a third entity that none of them could have produced alone.

The television series is the primary text. The online fan communities — message boards, wikis, podcasts, fan fiction archives — are the secondary text. But what the fans and producers create together, through the unintentional collaboration of their parallel activities, is a third text. This is what I called "the third world": the world produced neither by the producers nor by the fans, but by their interaction.

The producers seed Easter eggs into episodes. The fans discover, catalog, and interpret them. The producers, aware that fans are cataloging and interpreting, adjust future episodes in response — not by taking direction from fans, but by being aware of the interpretive context in which their work will be received. Carlton Cuse told me explicitly that the producers stayed away from the fan boards, in part because criticism "can really stick in your brain." Even avoiding the fan boards is a relationship with them. The avoidance shapes the work. The third world exists whether the producers participate in it or not.

The fans, meanwhile, were not simply consuming a text. Anthropologist Henry Jenkins had described what he called "textual poachers" – fans who take the raw material of a commercial text and repurpose it for their own cultural production. *Lost* fans wrote fan fiction extending the narrative, created elaborate theoretical frameworks explaining the island's mysteries, built wikis that collectively constituted the most comprehensive reference to the show that existed anywhere. Lostpedia, the community-created wiki, was more detailed and more accurate on many points than any official source. The fans had collectively become the definitive scholars of the text they loved.

In the theoretical framework I was working with in 2008, I drew on Abner Cohen's observation that "culture generally is expressed in terms of symbolic forms and performances that are by definition ambiguous, having reference to both political and existential issues simultaneously." Judith Irvine's work on shared language and community formation. Michael Silverstein and Greg Urban on text as the building block of cultural transmission.

These are academic tools for a real observation: the fan community of *Lost* had the structural characteristics of a culture. It had a shared mythology — the island, the numbers, the Dharma Initiative. It had shared language — the Jaters and Skaters, the game-changers and foilers, the Easter eggs and ARGs. It had rules and norms, enforced through community consensus. It had rituals — the weekly episode watch, the post-episode message board discussion, the podcast, the between-season speculation. It had, in some cases, physical gatherings that reinforced the community through face-to-face presence.

The community had formed around a commercial television product, but it was no longer reducible to that product. It had become something with its own internal logic, its own social structure, its own forms of production and meaning-making. The show could have been cancelled and the community would have continued, as in fact it did after the finale aired and the debate about the ending became, itself, an ongoing cultural event within the community.

I should be honest about what the third world was not.

The fans believed, many of them, that they had some hand in shaping the show's direction. The producers consistently said they did not monitor the fan boards and did not take direction from them. The evidence supports the producers' account. The show was, Cuse and Lindelof insisted, character-driven and pre-structured in its major arcs. Fan theories — however elaborate, however well-documented — did not determine what happened on the island.

The third world was a world of interpretation, not authorship. The fans were not co-writing *Lost*. They were writing their own parallel text alongside it — a running commentary, a shared interpretive framework, a community of practice organized around the act of watching. This is genuinely meaningful. It is also importantly different from having creative control.

The architecture of contingency gave fans the feeling of participation without the substance of it. The Easter eggs rewarded attention without providing genuine narrative input. The ARGs extended the story without changing it. The producers remained entirely in control of what happened on the island. The fans were engaged, active, creative, and consequential in the cultural meaning the show generated — while being structurally excluded from the decisions that shaped it.

This tension between the experience of participation and the reality of exclusion is not unique to *Lost*. It is the structural condition of fandom as such. The third world is real, and it matters, and it is not what the fans think it is. That, too, is part of its nature.

Chapter Nine: What Lost Started

Lost ended on May 23, 2010. The finale, "The End," resolved the character arcs with an afterlife framing that the producers had set up, quietly, across the final season. It did not explain the island's mechanism. It did not fully account for the mythology that six seasons of Easter eggs had constructed. It chose the emotional truth of the characters over the narrative completeness of the mythology — which was exactly the choice Cuse and Lindelof had always said they would make, and which a substantial portion of the fan community experienced as a betrayal.

The argument about the ending continues. That it continues is itself evidence of what the show built.

But the argument about the ending is not the legacy of *Lost*. The legacy is everywhere else.

Everything I observed in 2008 has become standard practice.

Marvel's connected cinematic universe — which began in 2008, the same year I was writing my thesis, with *Iron Man* — is the most successful deployment of transmedia architecture in history. The post-credits sequences that plant seeds for future films are Easter eggs. The tie-in Disney+ series that expand the backstory are ARGs in narrative form. The fan wikis, the speculation forums, the "Explain the Marvel Timeline" YouTube videos with tens of millions of views — these are the descendants of Lostpedia, built by a community that learned from the *Lost* playbook what engaged fandom looked like and what it required.

HBO's prestige drama model — *Game of Thrones*, *House of the Dragon*, *The Last of Us* — builds deliberately for the community of interpretation. "Book readers" and "show-only" viewers constitute distinct fan demographics with their own subcultures. Adaptation debates, episode discussions, theory crafting: the architecture of contingency has been standardized

Netflix built an entire business model around what *Lost* taught about binge viewing and serialized engagement. The completion anxiety — the need to know what happens next — that the architecture of contingency engineered into weekly broadcast television was weaponized into the autoplay function that begins the next episode before you have fully processed the last one. The four levels of fans Lindelof and Cuse designed for have become the four revenue segments that streaming platforms optimize for.

ARGs are now standard marketing. Viral campaigns, fake websites, Easter eggs planted in trailers, "hidden" announcements embedded in social media posts — every major franchise release has a transmedia component. The *Lost* Experience, which felt experimental in 2005, is now an industry standard deliverable with its own production budgets and specialist agencies.

When I described my thesis subject to people outside the anthropology department in 2008, the most common reaction was mild bafflement. Studying a fan community as a culture? Using anthropological fieldwork methods to analyze the ecosystem of a television show? This seemed, to many, like applying serious tools to frivolous material.

It was not, and the discipline has caught up. Digital ethnography — the anthropological study of online communities, practices, and cultural formations — is now an established subfield. The methodological questions I was working through in 2008 — how to conduct participant-observer research in a community distributed across multiple platforms and geographies, how to analyze a "text" that exists simultaneously in a television broadcast, on dozens of websites, in physical ARG objects, and in the ongoing discourse of its community — are questions that hundreds of researchers are now working to answer with rigorous tools.

The insight that fan communities have the structural characteristics of cultures — shared mythology, language, rules, rituals, and forms of production — is no longer controversial. It is the starting premise of an entire field. The question has moved from "can we study this?" to "how do we study this more precisely?"

The birth of digital anthropology as a practice — the systematic use of anthropological methods to study human communities that exist primarily or substantially online — coincides almost exactly with the period of *Lost*'s broadcast run. This is not coincidental. The show was large enough, and the community that formed around it rich enough, to make visible at scale what had previously been observable only in smaller online formations. *Lost* gave researchers a case study of unprecedented size and complexity. The field grew around it.

The Tropical Smoothie Cafe on Oahu, where forty fans gathered to watch the season four premiere in 2008, is gone. Not the cafe — the practice. The gathering of fans in a physical space to watch a television show together is essentially obsolete. Streaming eliminated the shared broadcast event. The COVID-19 pandemic accelerated a transition that was already underway. The "water cooler moment" — the shared cultural event that everyone experienced simultaneously and discussed together the next day — has been replaced by the algorithmically personalized feed, where everyone sees different content at different times and the shared frame of reference is increasingly fragmented.

What this means for the architecture of contingency is complex. On one hand, the tools available for building transmedia ecosystems are vastly more sophisticated in 2026 than in 2004. On the other, the shared broadcast moment that made the weekly episode of *Lost* a national event — that made the season three finale leak a genuine cultural catastrophe — no longer exists in quite the same form. You cannot have a shared spoiler panic about a show that everyone is watching on their own schedule.

Something was lost when *Lost* ended. Not just the show, but the specific cultural conditions in which the show had thrived: the weekly broadcast rhythm that created shared anticipation, the pre-social-media internet that concentrated fan discussion in specific forums rather than dispersing it across a thousand platforms, the moment before algorithmic content curation when a show could be a genuine shared cultural object rather than a niche recommendation targeted at a specific demographic segment.

The woman at the Tropical Smoothie Cafe who turned to me in shock and said "Was that a polar bear?" — that moment required a specific set of conditions. It required a shared screen. A shared time. A room full of strangers who had all just seen the same thing at the same moment and had no idea what it meant. The technology that followed *Lost* was built on what *Lost* taught, and in building on it, it dissolved some of the conditions that had made the original possible.

The island predicted the future. The future turned out to be more like the island than we expected: full of mysteries, contingent, shaped by forces we do not fully understand, and populated by people arguing about what it all means. We are still waiting for the finale.

Bibliography and Further Reading

Biressi, A. and Nunn, H. (2005). *Reality TV: Realism and Revelation.* Wallflower Press, London.

Brooker, Will (2002). *Using the Force: Creativity, Community, and Star Wars Fans.* Continuum, New York.

Baym, Nancy K. (2000). *Tune In, Log On: Soaps, Fandom, and Online Community.* Sage Publications, Newbury Park, CA.

Card, Orson Scott (2006). *Getting Lost: Survival, Baggage, and Starting Over in J.J. Abrams' Lost.* Benbella Books, Dallas.

Cohen, Abner (1980). "Drama and Politics in the Development of a London Carnival." *Man*, New Series, Vol. 15, No. 1, pp. 65–87.

Colson, Elizabeth (1979). "The Intensive Study of Small Sample Communities." *The Craft of Social Anthropology.* Pergamon Press, Oxford.

Defoe, Daniel (2001 [1719]). *Robinson Crusoe.* Modern Library, New York.

Fiske, J. (1992). "The Cultural Economy of Fandom." In Lewis, Lisa (Ed.), *The Adoring Audience*, pp. 30–49. Routledge, London.

Geertz, Clifford (1973). *The Interpretation of Cultures.* Basic Books, New York.

Gray, Jonathan and Mittell, Jason (2006). "Speculation on Spoilers: Lost Fandom, Narrative Consumption, and Rethinking Textuality." Unpublished manuscript.

Gwenllian-Jones, Sara (2004). "Virtual Reality and Cult Television." In Gwenllian-Jones, S. and Pearson, R. (Eds.), *Cult Television*, pp. 83–97. University of Minnesota Press, Minneapolis.

Hills, Matt (2005). "The Aleatory Object in Psychoanalytic Ethnography and Cyclical Fandom." *American Behavioral Scientist*, Vol. 48, No. 7, pp. 801–821.

Intintoli, Michael James (1984). *Taking Soaps Seriously: The World of Guiding Light*. Praeger Publishers, New York.

Irvine, Judith (1996). "Shadow Conversations: The Indeterminacy of Participant Roles." In Silverstein, M. and Urban, G. (Eds.), *Natural Histories of Discourse*, pp. 131–159. University of Chicago Press, Chicago.

Jenkins, Henry (1992). *Textual Poachers: Television Fans and Participatory Culture*. Routledge, New York.

Jenkins, Henry (2006). *Convergence Culture: Where Old and New Media Collide*. New York University Press, New York.

Kaveney, Roz (2005). *From Alien to the Matrix: Reading Science Fiction Film*. I.B. Tauris, London.

Lofland, John and Lofland, Lyn H. (2005). *Analyzing Social Settings: A Guide to Qualitative Observation and Analysis*. Wadsworth Publishing, Belmont, CA.

Powdermaker, Hortense (1951). *Hollywood: The Dream Factory — An Anthropologist Looks at the Movie-Makers*. Little, Brown, and Company, Boston.

Silverstein, Michael and Urban, Greg (1996). "The Natural History of Discourse." In *Natural Histories of Discourse*, pp. 1–17. University of Chicago Press, Chicago.

Van Velson, J. (1967). "The Extended Case Method and Situational Analysis." In Epstein, A.L. (Ed.), *The Craft of Social Anthropology*. Tavistock, London.

Vaz, Mark Cotta (2005). *The Lost Chronicles: The Official Companion Book*. Hyperion Books, New York.

Verne, Jules (2004 [1874]). *The Mysterious Island*. Modern Library, New York.

Jenkins, Henry, Ford, Sam, and Green, Joshua (2013). *Spreadable Media: Creating Value and Meaning in a Networked Culture*. NYU Press.

Mittell, Jason (2015). *Complex TV: The Poetics of Contemporary Television Storytelling*. NYU Press.

Booth, Paul (2010). *Digital Fandom: New Media Studies*. Peter Lang.

Hills, Matt (2002). *Fan Cultures*. Routledge.

Pearson, Roberta (2010). "Fandom in the Digital Era." *Popular Communication*, Vol. 8, No. 1.

About the Author

CD Damitio has spent the better part of three decades living, working, and wandering across four continents — from Morocco to Japan, Southeast Asia to the American West. He traded a TV and a VCR for a VW bus in 1996 and never quite looked back. His books cross genre lines the way he crosses borders: with curiosity, irreverence, and a stubborn refusal to settle for the official version of anything.

He is the founder of Baoism — a philosophy built around Bring Your Own Teachers, Bring Your Own Truth, and the radical idea that you are allowed to build your own world. Baoism is not a course, not a movement, and definitely not a guru product. It is a framework for people who are done waiting for permission. Explore it at Baoism.org.

His full catalog — fiction, memoir, philosophy, and everything in between — lives at Indignified.com/books. If you want to follow what he's thinking about, building, or arguing with himself over, find him on Xcrol.com/@cd, the privacy-first social network he built because the existing ones weren't worth the trade.

Made in the USA
Monee, IL
05 November 2026

67816240R00030